SOCIAL STUDIES MATTERS

I0801027

WHAT IS GEOGRAPHY?

PETER FINN

PowerKiDS press™

Published in 2024 by The Rosen Publishing Group, Inc.
2544 Clinton Street, Buffalo, NY 14224

Copyright © 2024 by The Rosen Publishing Group, Inc.

All rights reserved. No part of this book may be reproduced in any form without permission in writing from the publisher, except by a reviewer.

First Edition

Editor: Therese Shea
Book Design: Michael Flynn

Photo Credits: Cover, p. 1 Goinyk/Shutterstock.com; (series background) Eky Studio/Shutterstock.com; p. 5 YaromirM/Shutterstock.com; p. 7 Amanda Mohler/Shutterstock.com; p. 9 Jes2u.photo/Shutterstock.com; p. 11 Mateusz Lopuszynski/Shutterstock.com; p. 13 Aida Majoral/Shutterstock.com; p. 15 artjazz/Shutterstock.com; p. 17 (theodolite) Freedy/Shutterstock.com; p. 17 Emre Akkoyun/Shutterstock.com; p. 19 Bizi88/Shutterstock.com; p. 21 Scott Prokop/Shutterstock.com.

Library of Congress Cataloging-in-Publication Data

Names: Finn, Peter, 1978- author.
Title: What is geography? / Peter Finn.
Description: Buffalo, New York : PowerKids Press, [2024] | Series: Social studies matters! | Includes index.
Identifiers: LCCN 2023034399 (print) | LCCN 2023034400 (ebook) | ISBN 9781499443929 (library binding) | ISBN 9781499443912 (paperback) | ISBN 9781499443936 (ebook)
Subjects: LCSH: Geography–Juvenile literature.
Classification: LCC G133 .F52 2024 (print) | LCC G133 (ebook) | DDC 910–dc23/eng20230722
LC record available at https://lccn.loc.gov/2023034399
LC ebook record available at https://lccn.loc.gov/2023034400

Manufactured in the United States of America

Some of the images in this book illustrate individuals who are models. The depictions do not imply actual situations or events.

CPSIA Compliance Information: Batch #CWPK24. For Further Information contact Rosen Publishing at 1-800-237-9932.

Find us on

CONTENTS

Our Wonderful World

Earth has so many wonderful **features**. People love Earth's wide oceans, high mountains, green forests, and dry deserts. Are you interested in how these places came to be? If so, you're interested in geography. Geography is the study of Earth's **surface**.

Geology

People who study geography are called geographers. Some **focus** on geology. Geology is the study of how Earth formed, what it's made of, and its natural **processes**. For example, the Colorado River carved, or cut, the Grand Canyon over **millions** of years!

Weather, Water, and More

Other geographers study how Earth's features affect **climate**. For example, mountains can block rain. The water cycle is another focus of geography. This is the path that water takes as it moves around Earth. These things affect how people, animals, and plants live on Earth.

People Too

Some geographers study where people live on Earth. They study how people adapt, or change, to live in certain places. They also study how people's actions change Earth. For example, people build dams across rivers, dig mines in mountains, and cut down forests.

dam

Why It's Important

Geography helps us understand the world around us today and our place in it. Geographers also help us learn how Earth has changed and how it might change in the years to come. This **information** can help us plan for the future.

Knowing about geography helps us better use Earth's **natural resources**. For example, people in windy places can use **turbines** to produce electricity, or power. People in sunny places can build solar farms, which change sunlight to electricity.

Tools

Geographers use tools to help with their work. They use a tool called a theodolite to measure how far and how high features are. **Satellites** take pictures of Earth's surface from high up. This is a way to study Earth's larger features.

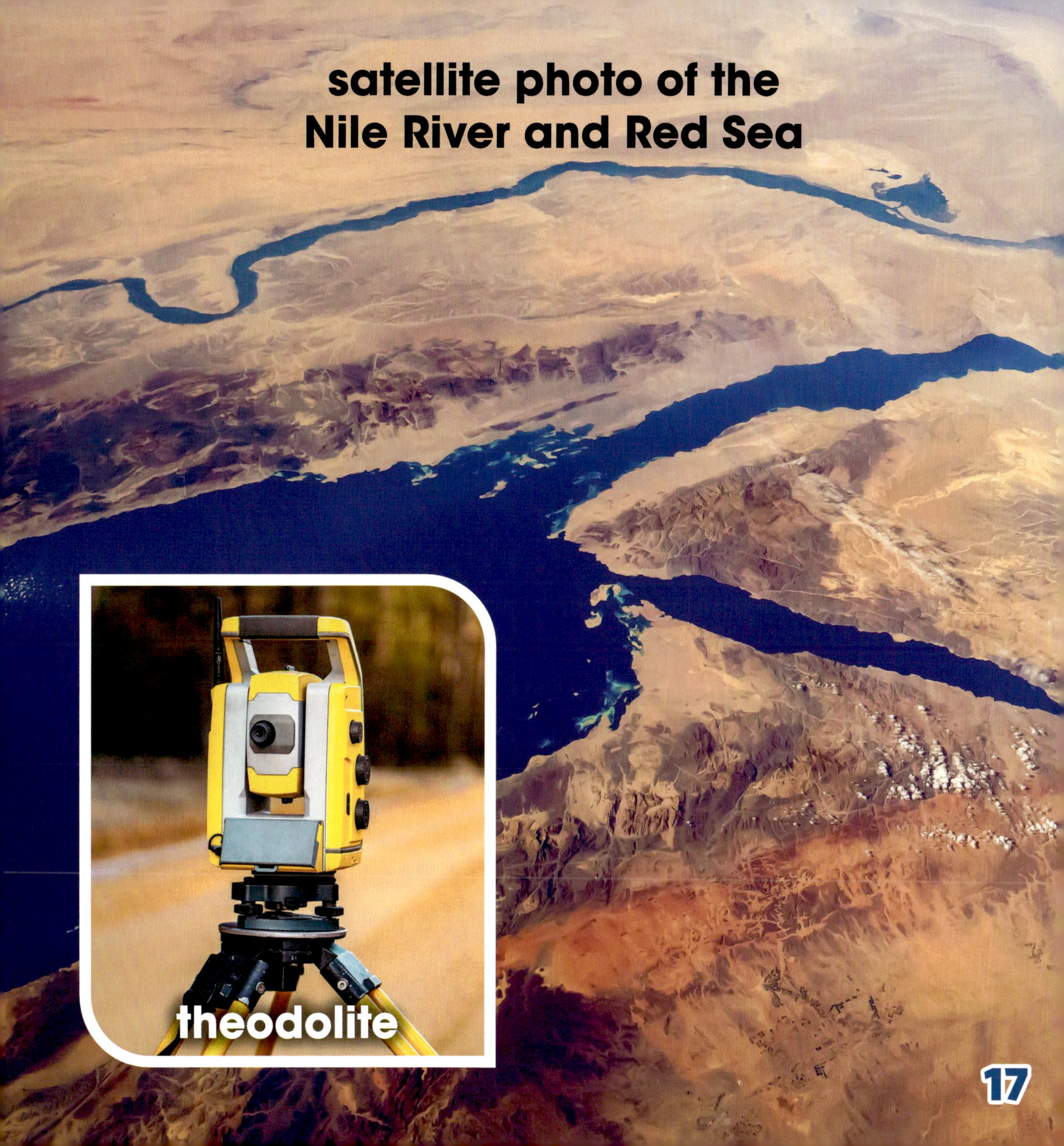

satellite photo of the Nile River and Red Sea

theodolite

Maps

Maps are geographers' tools too. Geographers who make maps are called cartographers. There are different kinds of maps. Political maps tell us where country borders, cities, and towns are. Topographic maps show natural land features, such as rivers, lakes, and hills.

We Need Geography

You might think every place on Earth has been mapped. That's not true. Only about 10 percent of the ocean floor has been mapped so far. Geographers are hard at work mapping the rest. Earth is always changing, so we'll always need geography!

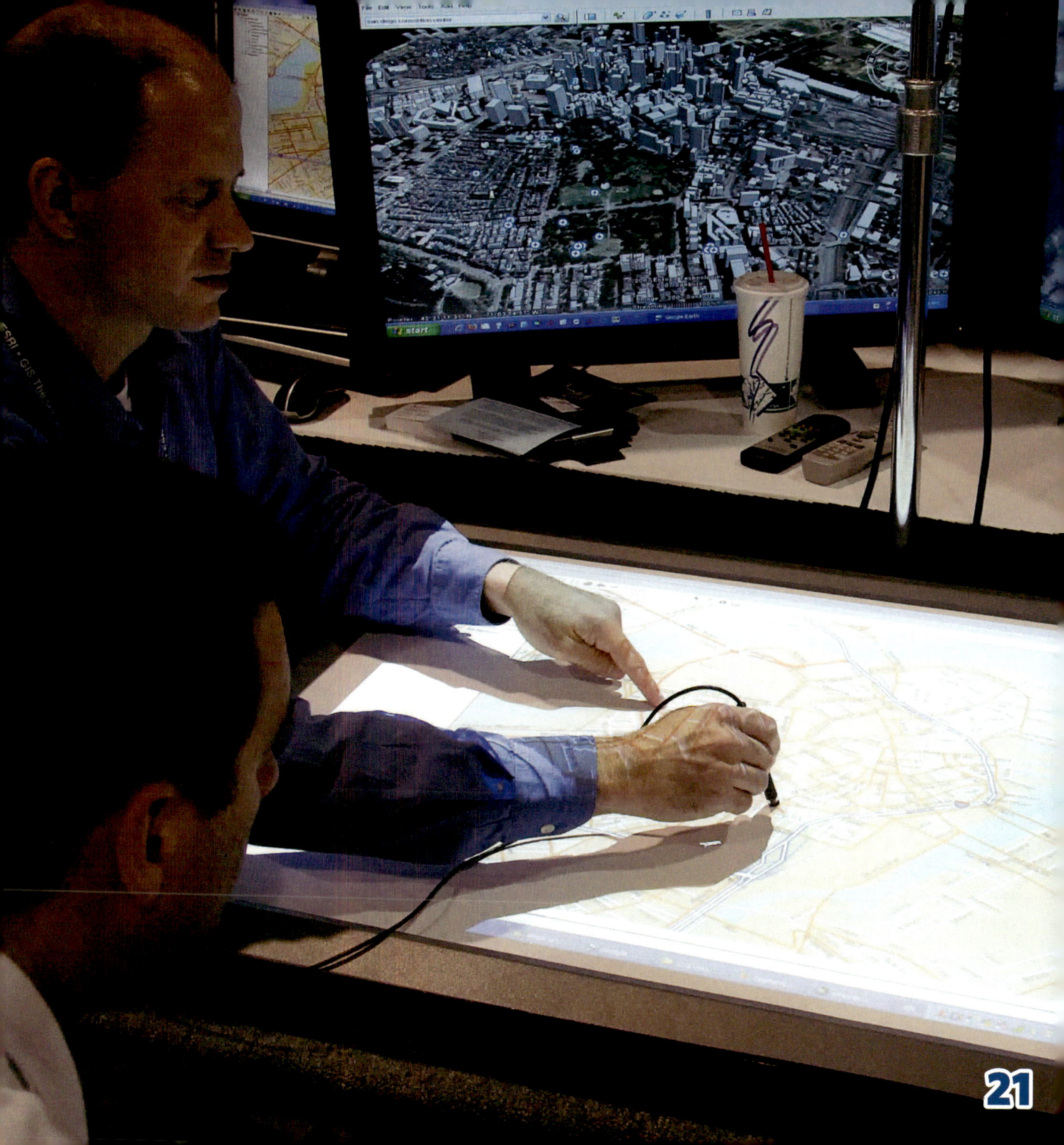

GLOSSARY

climate: The average weather conditions of a place over a period of time.

feature: An interesting or important part.

focus: To direct attention or effort to something.

information: Knowledge obtained from study or observation.

million: A thousand thousands, or 1,000,000.

natural resource: Something in nature that can be used by people.

process: A series of steps or actions before something is completed.

satellite: An object that circles Earth in order to collect and send information.

surface: The top layer of a planet; the ground.

turbine: A motor operated by the movement of water, steam, or air.

FOR MORE INFORMATION

BOOKS

Brinded, Alex. *Mapping Physical Geography*. New York, NY: KidHaven Publishing, 2020.

National Geographic Kids. *Beginner's World Atlas*. Washington, D.C.: National Geographic, 2022.

WEBSITES

Geography
www.ducksters.com/geography/
Learn about the field of geography on this site.

What Is Geography?
www.nationalgeographic.org/education/what-is-geography/
National Geographic has many exciting and interesting links.

Publisher's note to educators and parents: Our editors have carefully reviewed these websites to ensure that they are suitable for students. Many websites change frequently, however, and we cannot guarantee that a site's future contents will continue to meet our high standards of quality and educational value. Be advised that students should be closely supervised whenever they access the internet.

INDEX